ONE IN A MILLION

A First Book About Periods

This book is dedicated to my daughter and all of my students who inspire me tell this story over and over again until the world listens.

Author: Konika Ray Wong, M.Ed

Editor: Robin Katz
Illustrator: Mary Navarro / www.myjoyfulpages.com

All Rights Reserved. This book or any portion thereof may not be reproduced or used in any manner whatsoever without the express written permission of the publisher except for the use of brief quotations in a book review.
ISBN: 979-8-9881150-0-7

Copyright 2023 Girl Power Science, www.girlpowerscience.com

Publisher's Cataloging-in-Publication data
Names: Wong, Konika Ray, author. | Navarro, Mary, illustrator.
Title: One in a million : a first book about periods / by Konika Ray Wong, M.Ed; illustrated by Mary Navarro.
Description: San Francisco, CA: Girl Power Science, 2023. | Summary: In One in a Million, children meet an empowering uterus character that takes them on a journey that demystifies the process of ovulation and menstruation.
Identifiers: LCCN: 2023906763 | ISBN: 979-8-9881150-2-1 (hardcover) | 979-8-9881150-0-7 (paperback) | 979-8-9881150-1-4 (epub)
Subjects:LCSH Menstruation--Juvenile literature. | Menarche--Juvenile literature. | Teenage girls--Physiology--Juvenile literature. | Puberty--Juvenile literature. | BISAC JUVENILE NONFICTION / Girls & Women | JUVENILE NONFICTION / Health & Daily Living / Bodily Functions | JUVENILE NONFICTION / Social Topics / Adolescence | JUVENILE NONFICTION / Science & Nature / Anatomy & Physiology | JUVENILE NONFICTION / Health & Daily Living / Maturing
Classification: LCC RJ145 .W66 | DDC 612/.662--dc23

ONE IN A MILLION

A First Book About Periods

Konika Ray Wong, M.Ed

Illustrations by Mary Navarro

A NOTE FROM THE AUTHOR:
How to Use this Book

This book is intended to be a guided story for kids from age four and up to read with a grownup. It will give them an empowering first introduction to the concept of menstruation. I created this narrative for my own daughter. And throughout two decades of teaching kindergarten through grade six science, thousands of my students have heard these words.

Kids are naturally curious. And the younger they are, the more receptive and open-minded they are to new information. The age of pubertal onset is getting younger and younger as time goes on. Every school year, at least one of my third graders starts her period. The earlier we empower young students with knowledge about periods, the more it demystifies this healthy sign of growth. I suggest you nonchalantly read this book to your child or your students and then shelve it with your other books to show that this topic is no different.

The book is written as if speaking directly to a child with ovaries and a uterus who has yet to start their first period. It's a great book to introduce the concept of periods to all children, whether or not they have ovaries and a uterus. I used the phrase "kids with a uterus" so that this book would be inclusive. Females aren't the only menstruators. Trans men, gender nonbinary, and gender fluid folks might also have periods. I used the word "shero" in the book to include both of the gender pronouns "she" and "he."

The text is intentionally simple so that grownups deliver just the right amount of information without going into too much detail. After parents or teachers read it, questions will arise naturally.

I've included a list of the most common questions kids ask, with answers in my 'science teacher voice,' to give you a sense of the period-positive wording I use with my students.

"Kids are naturally curious. And the younger they are, the more receptive."

FREQUENTLY ASKED QUESTIONS:

1. **Does it hurt?**

   No. In fact, even grownups who have had their period for a long time often don't realize it has started until they notice it when they go to the bathroom.

2. **Do cramps hurt?**

   Cramps are rare when kids first start their period. You have probably experienced a muscle cramp if you've run around right after eating. So, you have a sense of what it feels like. The uterus is the strongest muscle in the body by weight. So, cramps can be uncomfortable. But there are easy ways to get relief. If you ever get cramps, just tell a grownup, and they will offer ways to help.

3. Does it get on your clothes?

There are many amazing inventions that protect your clothing. These include pads, period underwear, menstrual cups, and tampons. Maybe someday, you'll invent something even better!

4. When will I get my first period?

Kids with a uterus typically start their periods anytime between the ages of eight and fifteen. The two major clues that indicate your body is getting ready are breast buds and growth spurts. First periods often start two to three years after breast buds develop.

5. How long does it last?

Each period lasts three to seven days. When kids first start their period, the cycles are usually irregular until their brain gets more practice telling their ovaries what to do.

6. Do people with a uterus have periods for their whole lives?

No. The average age that people stop having their period is between forty-five and fifty-five. When periods stop, it is called 'menopause.'

7. Can people swim when they are having their period?

Yes. Some people wear period swimwear, and others wear tampons.

8. What happens to the other 999,999 eggs?

Once you start having your period, your body usually releases one egg each month. Many of the eggs break down and get absorbed by the ovaries. Kids with ovaries typically have between 300,000 to 500,000 eggs remaining when they start puberty.

"I used the word "shero" in the book to include both of the gender pronouns "she" and "he."

For more resources to empower you and your child to navigate puberty with confidence, visit girlpowerscience.com and follow us on Instagram @girlpowerscience and @konikraywong.

*While kindness and love figuratively flow from the heart and blood literally flows, the author intentionally took a departure from scientific accuracy. Blood typically elicits a squeamish response, and this poetic description attempts to curb that response.

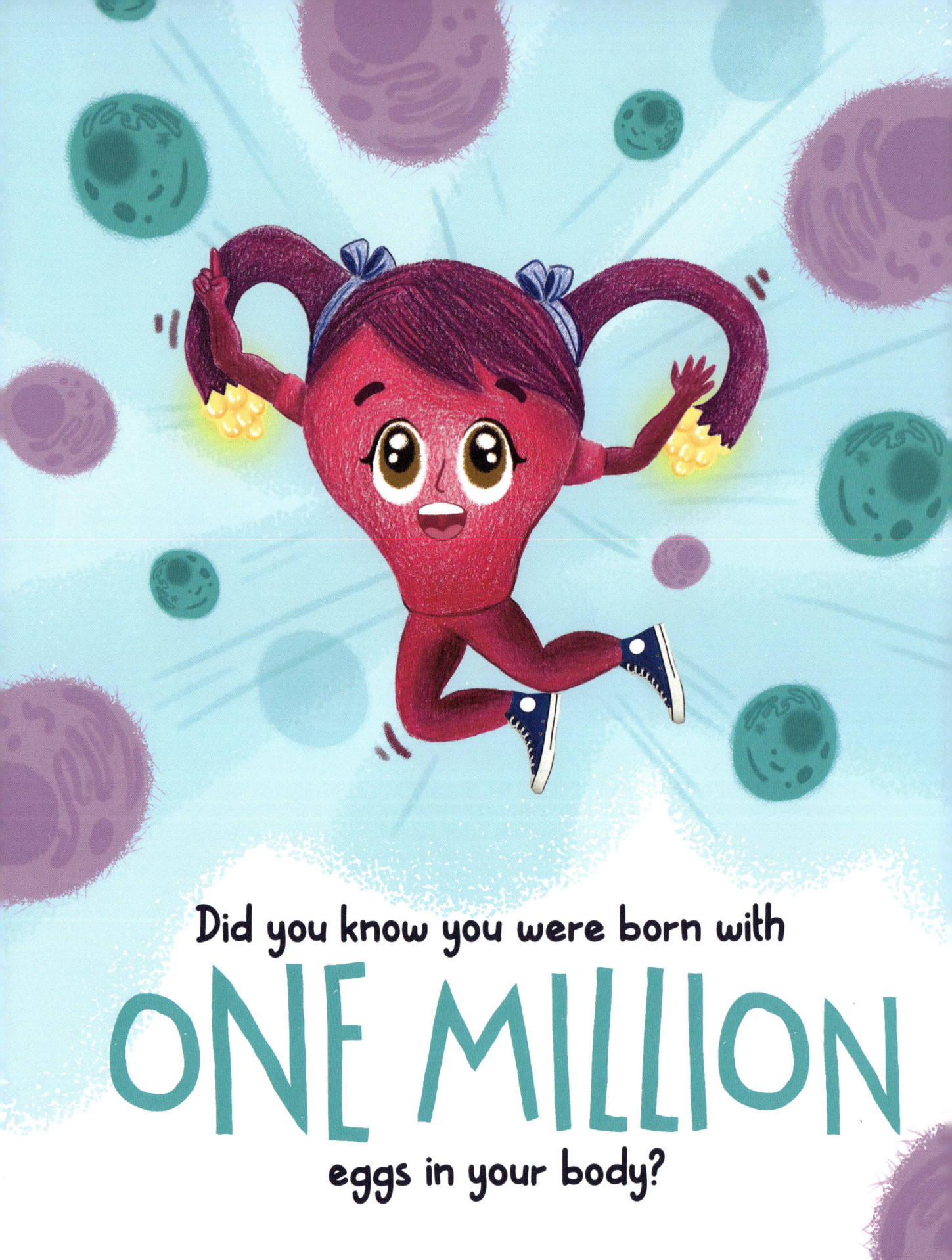

We don't lay eggs because we give birth to live babies, but

WE DO STORE EGGS IN OUR BODIES.

When we stand in the "super-shero" pose with our hands on our hips, our fingers point to where our eggs are stored!

Human eggs are tiny:
They're the size of pencil tips.

You have two ovaries:
One on each side of
your uterus. They
store your eggs.

Your ovaries are quite small:
The size of almonds.

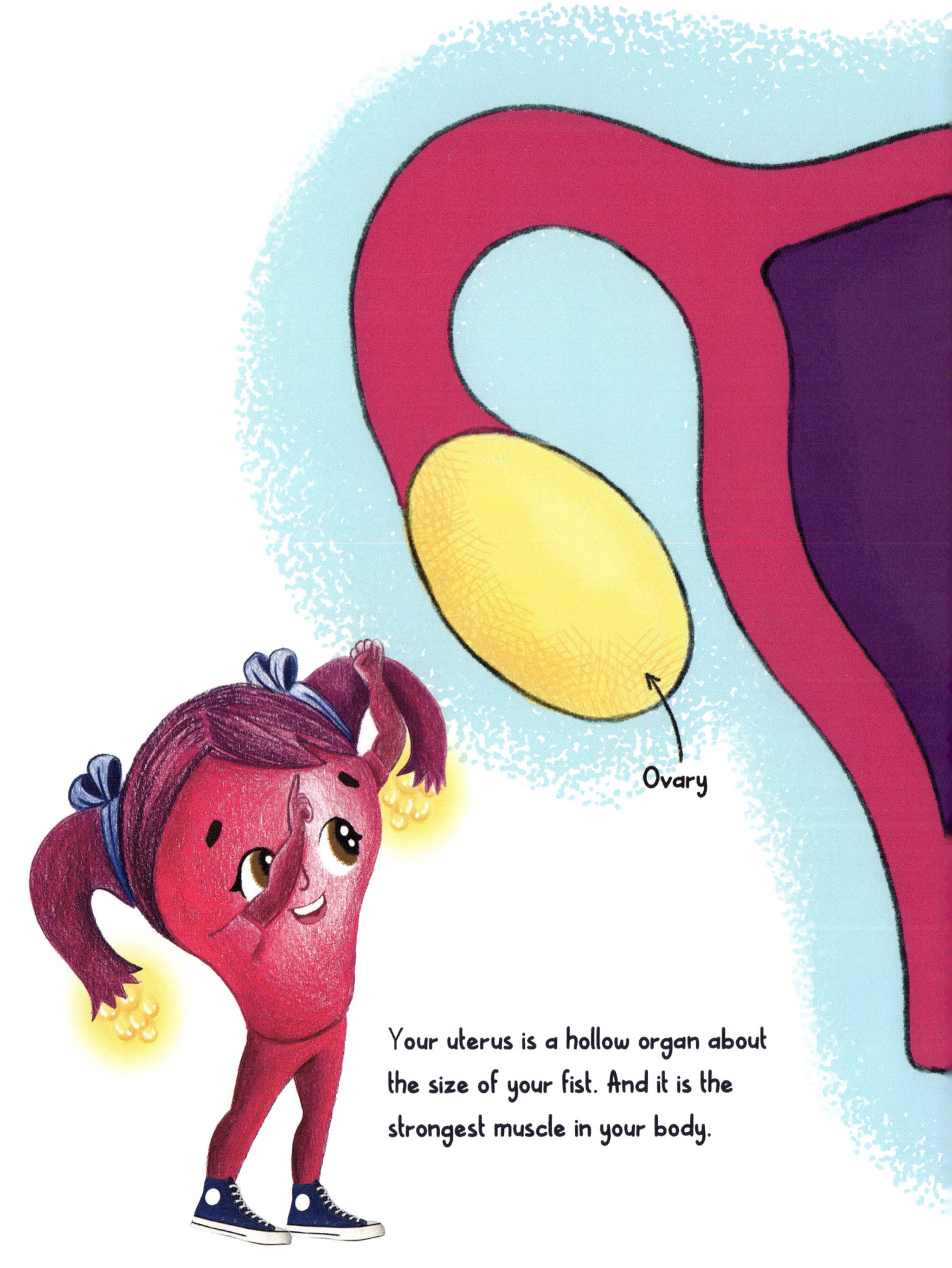

Ovary

Your uterus is a hollow organ about the size of your fist. And it is the strongest muscle in your body.

Uterus

A uterus is also called a womb. It is shaped like an upside down pear and located below your belly button.

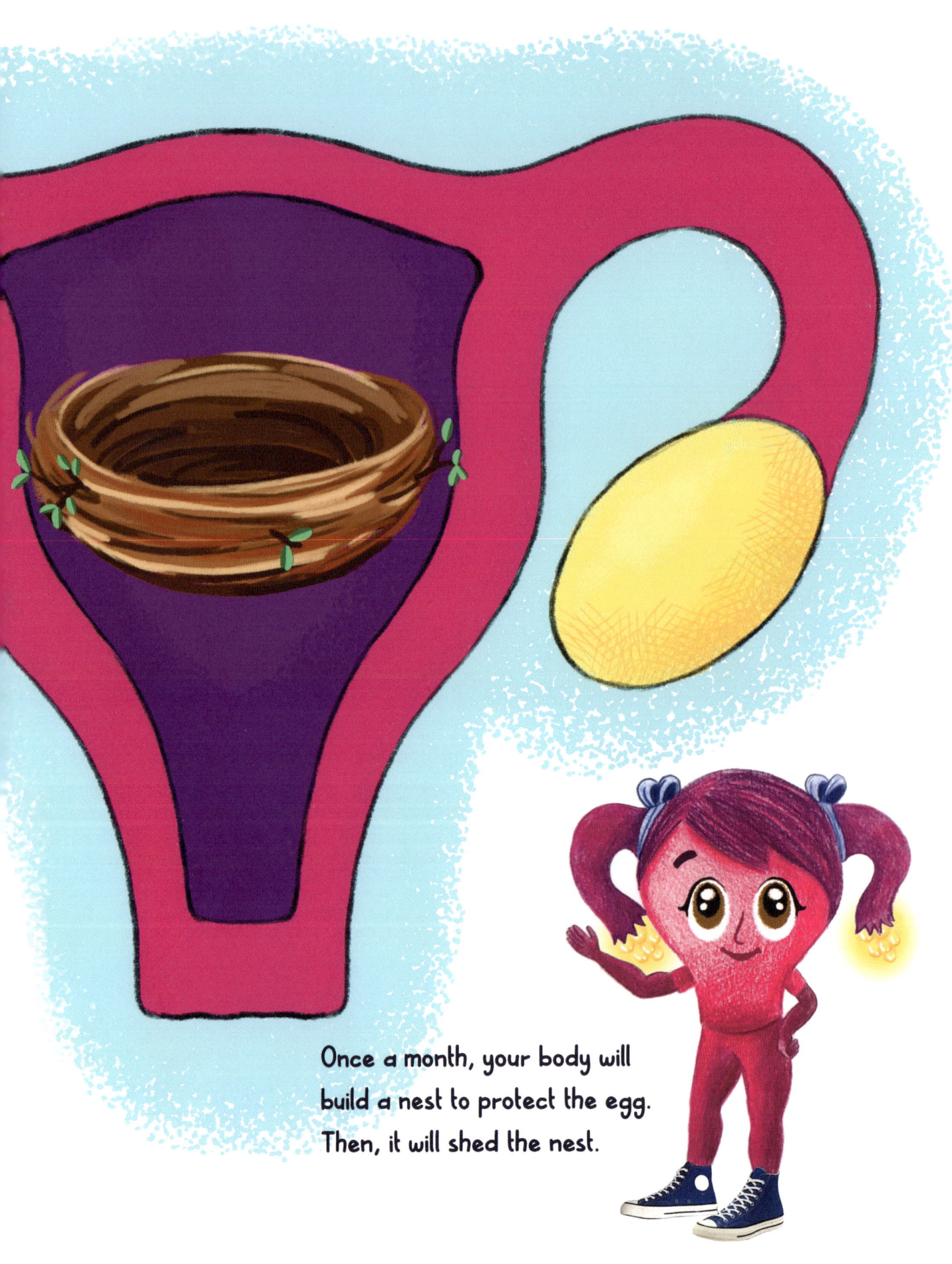

Once a month, your body will build a nest to protect the egg. Then, it will shed the nest.

But if someday you decide to have a baby, instead of shedding the nest, your body will keep it as a helpful place for the baby to grow.

I bet you are probably wondering what the nest is made of. It isn't made out of twigs like a bird's nest. Could you guess what it's made of if I give you four clues? Here they are:

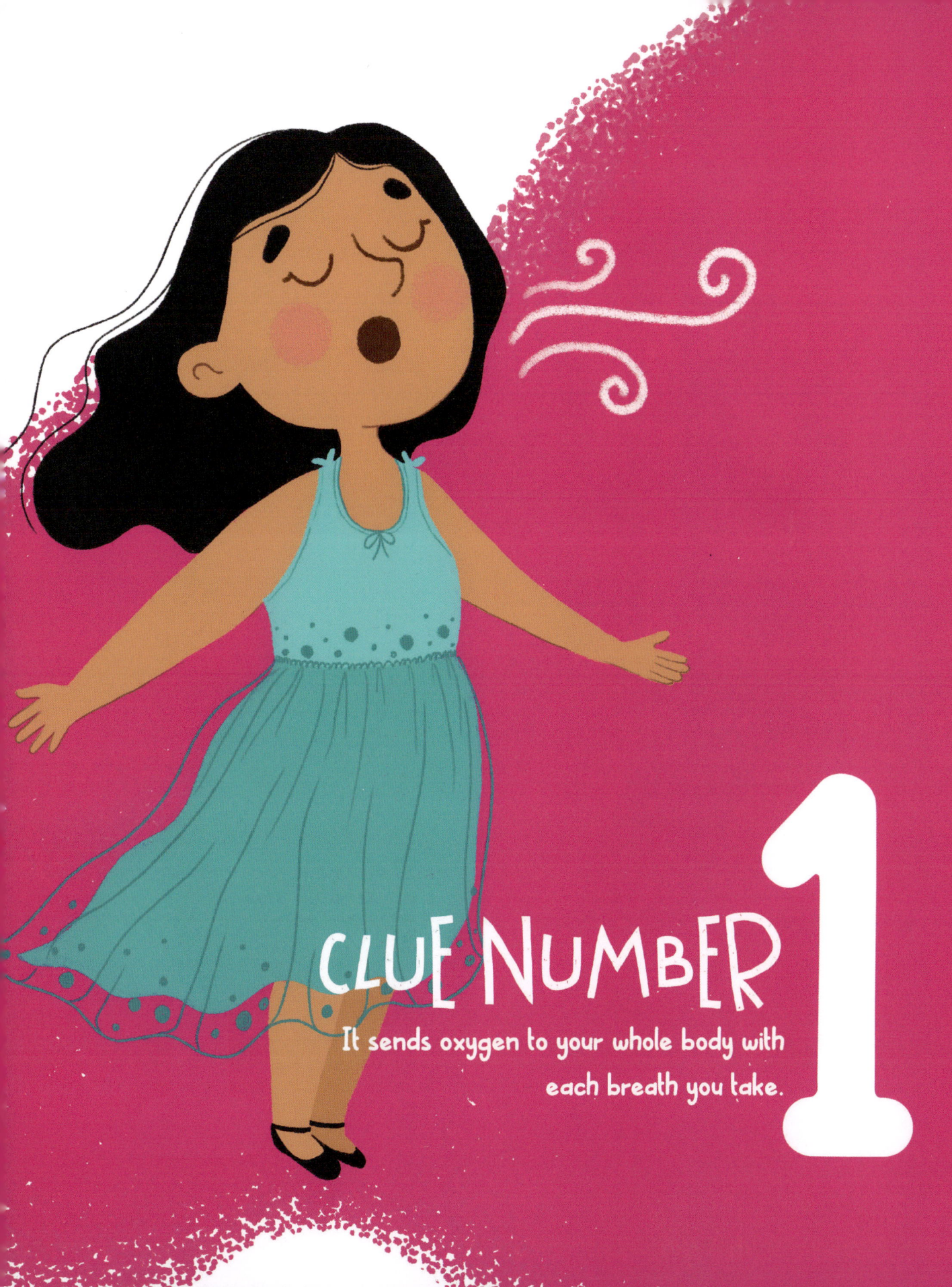

CLUE NUMBER 1

It sends oxygen to your whole body with each breath you take.

CLUE NUMBER 2

It fights germs like a brave ninja warrior.

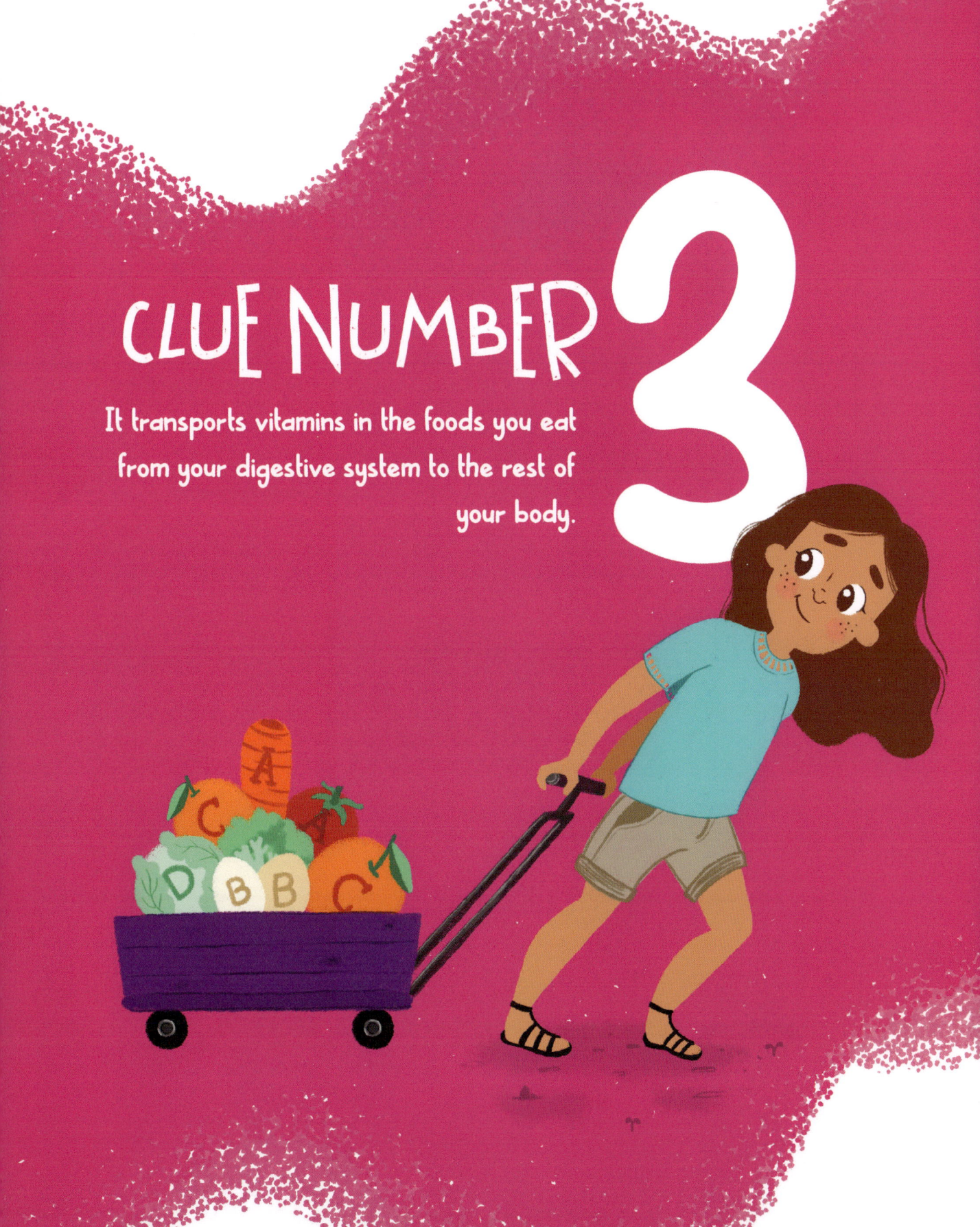

CLUE NUMBER 3

It transports vitamins in the foods you eat from your digestive system to the rest of your body.

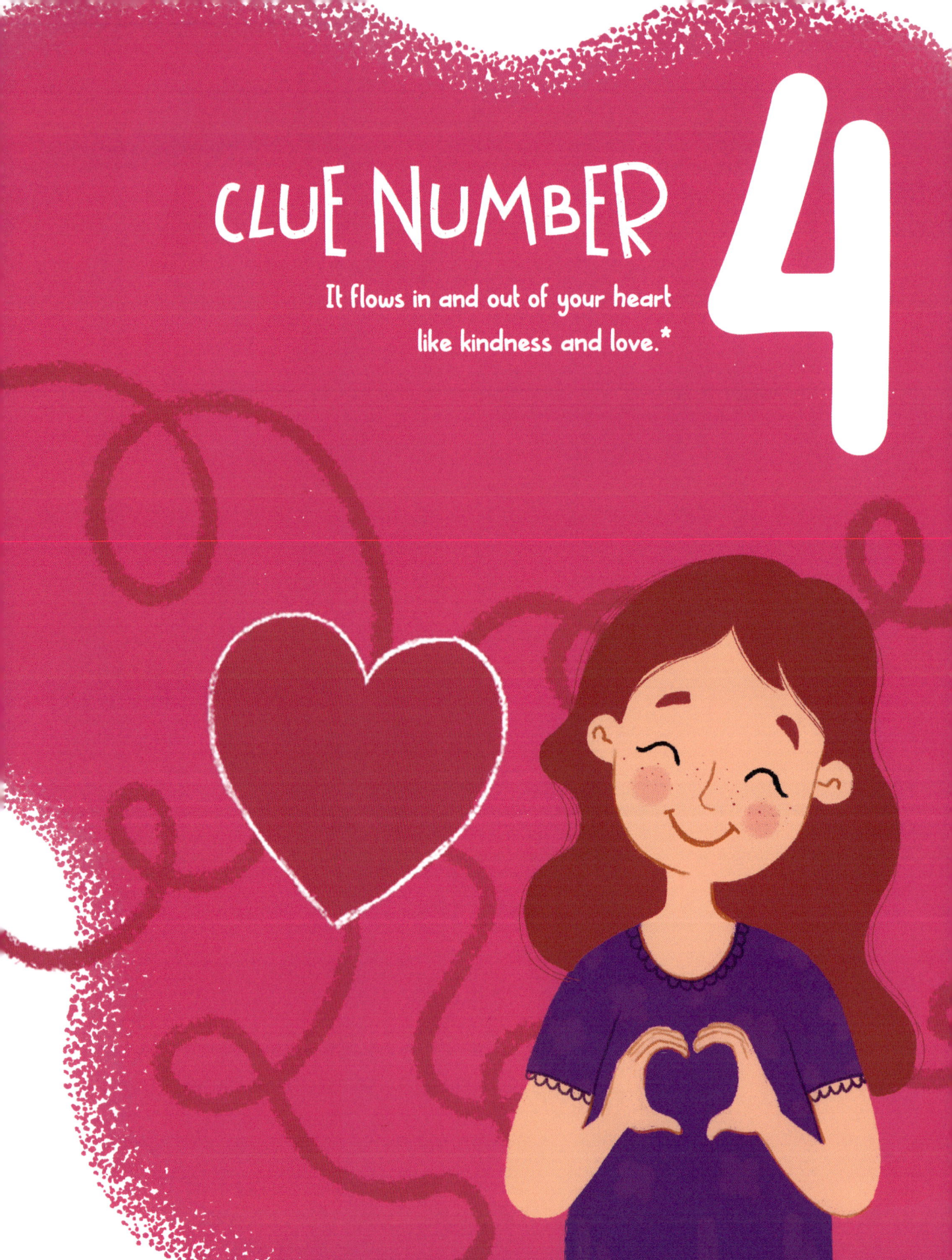

CLUE NUMBER 4

It flows in and out of your heart like kindness and love.*

The answer is...

BLOOD!

The four clues mention some of the most important superpowers of blood that help keep us healthy!

The nest in your uterus is made out of nutrient-rich blood. And every time the uterus sheds that blood, it's called 'having a period.'

Do you remember four ways that your

BLOOD KEEPS YOU HEALTHY?

(Hint: Recall the clues you read earlier!) The four clues mention some of the most important superpowers of blood that help keep us healthy!

Your grownups are so grateful that your body is healthy and growing!

Printed in Great Britain
by Amazon